Module 33209-05

Introduction to Codes and Standards

COURSE MAP

This course map shows all of the modules in the second level of the *Electronic Systems Technician* curriculum. The suggested training order begins at the bottom and proceeds up. Skill levels increase as you advance on the course map. The local Training Program Sponsor may adjust the training order.

ELECTRONIC SYSTEMS TECHNICIAN LEVEL TWO

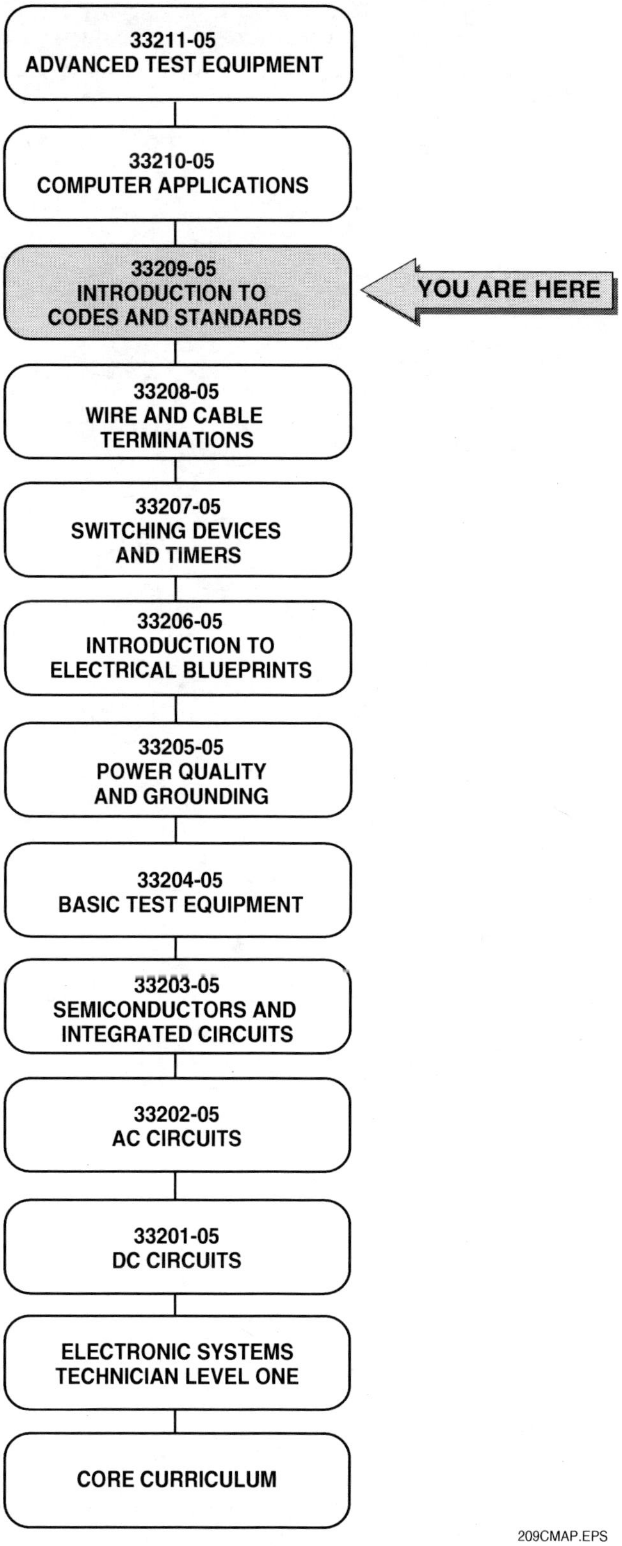

209CMAP.EPS

MODULE 33209-05 CONTENTS

Figures

Tables

MODULE 33209-05

Introduction to Codes and Standards

Objectives

When you have completed this module, you will be able to do the following:

1. Identify the difference between codes and standards.
2. Identify trade-relevant codes and standards and their applications.
3. Explain how use and find information in the *National Electrical Code® (NEC®).*
4. Explain the role of testing laboratories.
5. Use the NEC® to determine the specific requirements for a given telecommunications and/or life safety system application.
6. Use the applicable ANSI/TIA/EIA standards to determine the specific requirements for a given telecommunications and/or life safety system application.

Prerequisites

Before you begin this module, it is recommended that you successfully complete *Core Curriculum; Electronic Systems Technician Level One;* and *Electronic Systems Technician Level Two,* Modules 33201-05 through 33206-05.

Required Trainee Materials

1. Pencil and paper
2. Appropriate personal protective equipment
3. Copy of the latest edition of the *National Electrical Code®*

1.0.0 ◆ INTRODUCTION

Most of the work done by ESTs in telecommunications, life safety, security, and other systems is governed by industry standards and building codes. This module introduces and provides a brief overview of the scope and content for the major codes and standards that apply to the industry with an emphasis placed on the *National Electrical Code®* (NEC®). It is important that you become familiar with the codes and standards that govern your work. During the remainder of your training, you will use these documents extensively and it should become second nature for you to refer to them in the course of your work. A good rule of thumb is: when in doubt, look it up.

2.0.0 ◆ THE PURPOSE OF CODES AND STANDARDS

Building codes and standards regulate all construction work performed in North America. Used together, codes and standards serve to protect life, health, and property and to ensure the quality of construction.

2.1.0 Codes

Building codes and other codes normally address the minimum safety requirements that must be adhered to (*Figure 1*). They typically describe circumstances under which the use of a given type of system, equipment, or component is required, depending on the use and occupancy of a building or structure. Generally, codes are written so

Note: The designations "*National Electrical Code®*" and "NEC®," where used in this document, refer to the *National Electrical Code®*, which is a registered trademark of the National Fire Protection Association, Quincy, MA. *All National Electrical Code® (NEC®) references in this module refer to the 2002 edition of the NEC®.*

Figure 1 ◆ Codes.

that they can easily be adopted into law. Some examples of codes developed by the **National Fire Protection Association (NFPA)** and commonly used by ESTs include the following:

- *NFPA 70, The National Electrical Code®*
- *NFPA 72, The National Fire Alarm Code®*
- *NFPA 101, The Life Safety Code®*

Three of the following codes were formerly published and maintained nationally as accepted building codes. Even though they are no longer published, they still may be in force in some state and local jurisdictions.

- *International Building Code®*
- *NFPA 5000, Building and Construction Safety Code™*
- *National Building Code* (no longer published)
- *Uniform Building Code™* (no longer published)
- *Standard Building Code* (no longer published)

There are now only two active building codes: the *International Building Code®* and *NFPA 5000.* State and local governments may also have additional requirements that must be met. More information about these nationally accepted building codes is given later in this module. It is essential that everyone involved in the installation of electronic systems be familiar with the local and national codes in force in the locality where the systems will be installed.

2.2.0 Standards

Standards detail how the protection specified by the code is to be achieved (*Figure 2*). Performance standards specify functions and capabilities of hardware and conditions under which the equipment must operate. Standards should not be confused with technical specifications. For example, NFPA standards require that an audible fire alarm signal be "clearly heard throughout the building," which is usually defined as 15 dBA above ambient

Figure 2 ◆ Standards.

sound levels. It does not specify how many audible alarms to use or where to locate them to achieve the requirement. This kind of information would normally be given in a technical specification.

2.2.1 Types of Standards

There are many types of standards (*Figure 3*). The more common types are defined as follows:

- *International standards* – Mainly used to enhance trade, these standards use a common language and establish equivalent measurements.
- *National standards* – These standards usually define a method of manufacture or design. Two organizations that establish national standards are the **American National Standards Institute (ANSI)** and **National Electrical Manufacturers Association (NEMA)**.
- *Regional and local standards* – Both regional and local standards establish requirements that conform to state and local needs.
- *Safety standards* – These standards establish requirements related to safety issues.
- *Installation standards* – These standards establish requirements related to how the installation of systems and equipment is to be achieved.

Figure 3 ◆ Types of standards.

- *Industry standards* – Standards developed over the years that form the foundation of how most companies in an industry deal with complexities relating to systems and devices are called industry standards. For example, **Underwriters Laboratories (UL)** has developed standards on most aspects of the telecommunications and electronic security industries, including equipment, installation, and maintenance procedures. To be listed by UL, a company must demonstrate that its products meet the standards for each category.
- *Company standards* – Most companies have developed specific procedures for handling many aspects of your job. These procedures are usually based on applicable codes and standards as well as other good practice methods.
- *Manufacturer's instructions* – Instructions that come with the equipment are called manufacturer's instructions, and they are part of the approval or listing. If the instructions are not followed, the approval or listing is void.

2.2.2 Adopted and Recognized Standards

Adopted standards are standards that have been adopted through statutory or administrative laws in a specific geographic area. Once a document is adopted by reference in a state statute or local ordinance, it has the same force and effect as if printed therein in its entirety.

Recognized standards are standards that are generally accepted as relevant in civil litigation, even in areas where no standards have been formally adopted. Where no applicable code or standard has been officially implemented, it is common for the courts to compare actual circumstances in a case with nationally recognized standards.

The use of standards reduces liability. In civil trials, adopted standards carry the most weight. Where there is no adopted standard, the courts will frequently focus on the issue of compliance with recognized standards. In the case of a conflict between recognized standards and common industry practice, recognized standards will prevail. For example, an alarm system found to be in violation of the recognized standards after a fatal fire may create a liability for the installer.

In a jurisdiction where no particular standard has been legally adopted or no standard is in force, national standards may apply. In addition to the moral and social responsibilities associated with conducting your work correctly, there are added civil liability implications. Complying with or exceeding the recognized codes and standards is your best protection from liability. When faced with justifying an action in a court proceeding, a company is better off if they can cite recognized standards instead of a common company or industry practice.

3.0.0 ◆ DETERMINING WHICH CODES AND STANDARDS TO FOLLOW

There are many types of codes and standards that can apply to a particular system or job. The decision as to which codes and standards apply is made by the Authority Having Jurisdiction (AHJ).

Each job has at least one AHJ. The AHJ is the person or agency that decides the acceptability of the systems and devices being installed in accordance with code requirements.

The AHJ can be a one or all of the following: a building inspector; a fire code inspector; another federal, state, local, or regional authority; a designated insurance company representative; or the property owner. The AHJ determines which rules apply and ultimately resolves any conflicts between any conflicting codes or rules. The AHJ is the organization, office, or individual responsible for approving equipment, an installation, or a procedure.

Deciding which codes and standards apply is not always easy; it requires a familiarity with all of the following:

- Adopted national codes and standards
- Adopted regional codes and standards
- Adopted state codes and standards
- Adopted local codes and standards
- Adopted customer codes and standards
- Codes and standards adopted by your company
- Recognized national codes and standards
- Recognized regional codes and standards
- Recognized state codes and standards
- Recognized local codes and standards

As previously stated, codes and standards establish minimum requirements, thereby allowing work to exceed the requirements and still be considered in compliance. This does not mean that the code or standard requirements can arbitrarily be modified by an individual code enforcement official. Only a person or agency that serves as an AHJ is permitted to modify or supplement the published requirements; however, this should be done in writing through ordinances, statutes, and administrative laws in accordance with applicable legal requirements.

The AHJ is usually authorized to modify the requirements for existing buildings where strict application of the code is not practicable. The AHJ is generally permitted by code to accept an alternate method of compliance if, in the AHJ's judgement, it provides an acceptable level of safety. For example, if an old fire alarm system had 40 manual pull stations mounted at 66 inches above the floor and the current NFPA standard specifies a maximum mounting height of 54 inches, the benefit of bringing the pull stations up to code would not merit the cost.

4.0.0 ◆ WORDS WITH SPECIAL MEANINGS USED IN CODES AND STANDARDS

When interpreting codes and standards, certain words and terms have a special meaning and a major impact on how tasks are to be accomplished. The definition of these words follows:

- *Shall* – This is a mandatory requirement.
- *Should* – This is a recommendation or something that is advised but not required. Note that in some codes and standards the words *may* or *desirable* may be used interchangeably with the word *should*.
- *Approved* – When something is approved, such as an item of equipment, it is sanctioned, certified, and supported.
- *Listed* – When something is listed, such as an item of equipment, the listing person or body has added the equipment to a list, indicating that it shares certain qualities or meets certain tests with other similar listed equipment.
- *Accepted* – When something is accepted, it is believed to satisfy a requirement or standard.
- *Equipment listed for the application* – When used, this term requires that equipment used in a system be properly listed for the application. The key words are *listed for the application*. It is a common misunderstanding for people to assume that, when an item is "listed," it is listed for all applications. Many devices are listed; however, they may not be listed for the specific application. In the listing process, a device becomes listed under one or more of the standards, and should therefore meet the performance requirements of that (those) standard(s) if properly installed, tested, and maintained. It is important to understand which applications each listing is for and to only use products listed for that application.
- *Equivalent* – The "or equivalent" clause in most standards allows the use of systems, methods, or devices of equal or better quality, as long as sufficient documentation is provided to the AHJ to demonstrate the equivalency.

5.0.0 ◆ CODE DEVIATIONS AND CONFLICTS

Code deviations that substitute or provide an alternative feature as accepted by the AHJ are normally considered as conforming to code. Generally, acceptance of any measure not in complete compliance with applicable codes and standards should be specifically stipulated by the AHJ in writing. In the event of conflict between two or more standards, the more stringent requirement should be applied. In one sense, the only true conflict would be a situation in which there is no way to comply with multiple provisions at the same time. For example, if a building code called for manual fire stations to be mounted at 36 to 56 inches above the floor, and another standard specified 42 to 60 inches above the floor, this would appear to be a conflict. Because it is possible to comply with both requirements by installing the manual stations at 42 to 56 inches above the floor, it is not a true conflict.

6.0.0 ◆ *NATIONAL ELECTRICAL CODE®*

NFPA sponsors, controls, and publishes *NFPA 70*, the *National Electrical Code®* (NEC®). It is revised every three years. The NEC® is the most widely used and generally accepted code in the world. It has been translated into several languages. The primary purpose of the NEC® is the practical safeguarding of persons and property from hazards arising from the use of electricity. This purpose is stated in ***NEC Section 90.1***. Most federal, state, and local municipalities in the United States have adopted the NEC®, in whole or in part, as their legal electrical code. The NEC® establishes the minimum requirements for the installation of electrical equipment. Some states or localities adopt the minimum requirements established by the NEC®, then add more stringent requirements. Compliance with NEC® requirements increases the safety of electrical installations, which is the reason the NEC® is so widely used. The NEC® is used by many people, such as those listed here:

- Electricians and installers
- Lawyers and insurance companies to determine liability
- Fire marshals and electrical inspectors in loss prevention and safety enforcement
- Designers to ensure a compliant installation

THINK ABOUT IT

Code Changes

This photograph shows the first code book and the 2002 code book. Code changes occur every three years. Who can suggest changes to the *National Electrical Code®*? What might be reasons for submitting changes?

209SA01.EPS

A thorough knowledge of the NEC® is one of the first requirements for becoming a trained electronic systems technician. Although the NEC® itself states "This Code is not intended as a design specification or an instruction manual for untrained persons," it does provide a sound basis for the study of electrical installation procedures, under proper guidance. The NEC® has become the standard reference of the electrical construction industry. Anyone involved in electrical work should obtain an up-to-date copy and refer to it frequently.

Whether you are installing new systems or equipment, or retrofitting existing ones, all electrical work must comply with the current NEC® and all local ordinances. Like most laws, the NEC® is easier to work with once you understand the language and know where to look for the information you need.

6.1.0 The Layout of the NEC®

The NEC® begins with a brief history. The main text of the NEC® is organized into chapters, articles, parts, and sections. Annexes are provided at the back of the book to provide supporting information. *Figure 4* shows how the NEC® is organized.

6.1.1 Types of Rules

There are two basic types of rules in the NEC®: mandatory rules and permissive rules. It is important to understand these rules as they are defined in ***NEC Section 90.5***. Mandatory rules contain the words *shall* or *shall not* and must be adhered to. Permissive rules identify actions that are allowed but not required and typically cover options or

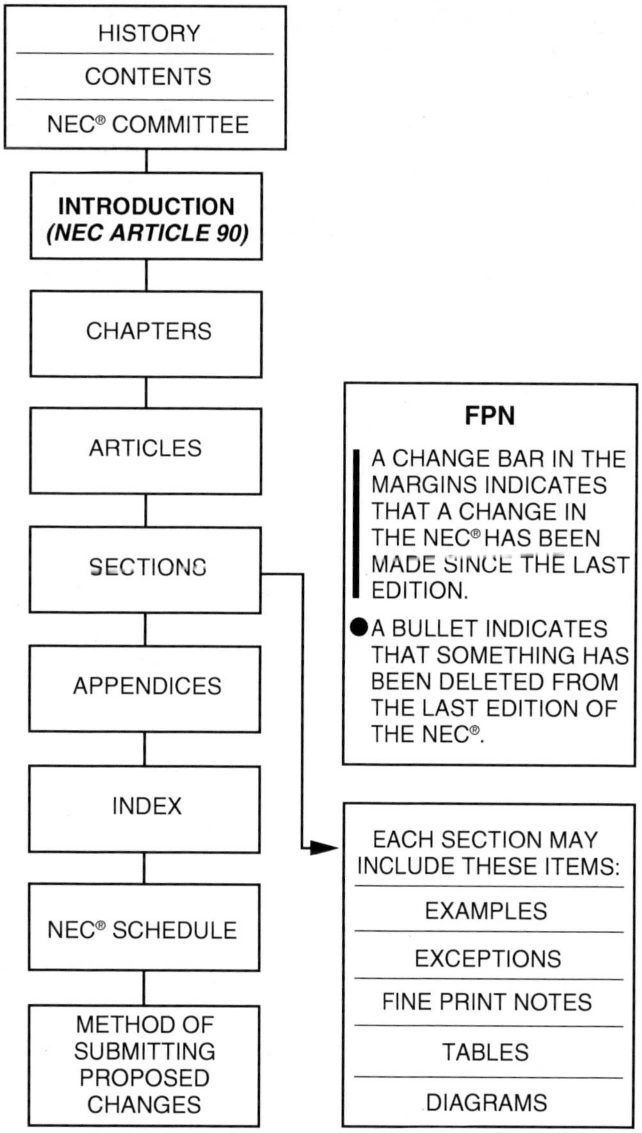

209F04.EPS

Figure 4 ◆ The layout of the NEC®.

INSIDE TRACK

NEC® Layout

Remember, chapters contain a group of articles relating to a broad category. An article is a specific subject within that category, such as ***NEC Article 250***, Grounding, which is in Chapter 2 relating to wiring and protection. When an article applies to different installations in the same category, it will be divided into parts using roman numerals (for example, ***NEC Article 250*** is divided into ten parts). Any specific requirements in any of the articles may also have exception(s) to the main rules.

The NEC® as a Reference Tool

The NEC®, although carefully organized, is a highly technical document; as such, it uses many technical words and phrases. It's crucial for you to understand their meaning. If you don't, you'll have trouble understanding the rules themselves. When using the NEC® as a reference tool, you may have to refer to a number of articles to find your answer(s). NEC® definitions are covered in ***NEC Article 100***. Many issues concerning the NEC® may be resolved by simply reviewing the definitions.

alternative methods. Be aware that local codes may amend requirements of the NEC®. This means that a city or county may have additional requirements or prohibitions that must be followed in that jurisdiction.

6.1.2 NEC® Introduction

The main body of the text begins with an *Introduction*, also entitled ***NEC Article 90***. This introduction gives you an overview of the NEC®. Items included in this section are as follows:

- Purpose of the NEC®
- Scope of the code book
- Code arrangement
- Code enforcement
- Mandatory rules and explanatory material
- Formal interpretation
- Examination of equipment for safety
- Wiring planning
- Metric units of measurement

6.2.0 The Body of the NEC®

The remainder of the book is organized into nine chapters. ***NEC Chapter 1*** contains a list of definitions used in the NEC®. These definitions are referred to as ***NEC Article 100***. ***NEC Article 110*** gives the general requirements for electrical installations. It is important for you to be familiar with this general information and the definitions.

6.2.1 NEC® Definitions

There are many definitions included in ***NEC Article 100***. You should become familiar with the definitions. You should become especially familiar with the following two definitions:

- *Labeled* – "Equipment or materials to which has been attached a label or other identifying mark of an organization that is acceptable to the authority having jurisdiction and concerned with product evaluation, that maintains periodic inspection of production of labeled equipment or materials, and by whose labeling the manufacturer indicates compliance with appropriate standards or performance in a specified manner."
- *Listed* – "Equipment or materials included in a list published by an organization acceptable to the authority having jurisdiction and concerned with product evaluation, that maintains periodic inspection of production of listed equipment or materials, and whose listing states either that the equipment or material meets appropriate designated standards or has been tested and found suitable for use in a specified manner."

Besides installation rules, you will also have to be concerned with the type and quality of materials that are used in electrical wiring systems. Nationally recognized testing laboratories are product safety certification laboratories. Underwriters Laboratories, Inc., also called UL, is one such laboratory. These laboratories establish and operate product safety certification programs to make sure that items produced under the service are safeguarded against reasonably foreseeable risks. Some of these organizations maintain a worldwide network of field representatives who make unannounced visits to manufacturing facilities to counter-check products bearing their seal of approval. The UL label is shown in *Figure 5*.

209F05.EPS

Figure 5 ◈ Underwriters Laboratories, Inc., label.

6.3.0 The Reference Portion of the NEC®

The annexes included with the NEC® provide reference sources. These reference sources can be used to determine the proper application of the NEC® requirements.

6.3.1 Organization of the Chapters

NEC Chapters 1 through 8 each contain numerous articles. Each chapter focuses on a general category of electrical application, such as ***NEC Chapter 2***, *Wiring and Protection*. Each article emphasizes a more specific part of that category, such as ***NEC Article 210***, *Branch Circuits, Part I General Provisions*. Each section gives examples of a specific application of the NEC®, such as ***NEC Section 210.4***, *Multiwire Branch Circuits*. ***NEC Chapter 9*** contains tables that are used when referenced by any of the articles in ***NEC Chapters 1 through 7***. ***Annexes A through F*** provide informational material and examples that are helpful when applying NEC® requirements.

The chapters of the NEC® are organized into four major categories:

- ***NEC Chapters 1, 2, 3, and 4*** – The first four chapters present the rules for the design and installation of electrical systems. They generally apply to all electrical installations.
- ***NEC Chapters 5, 6, and 7*** – These chapters are concerned with special occupancies, equipment, and conditions. Rules in these chapters may modify or amend those in the first four chapters.
- ***NEC Chapter 8*** – This chapter covers communications systems, such as the telephone and telegraph, as well as radio and television receiving equipment. It may also reference other articles, such as the installation of grounding electrode conductor connections as covered in ***NEC Section 250.52***.
- ***NEC Chapter 9*** – This chapter contains tables that are applicable when referenced by other chapters in the NEC®.
- ***Annexes A through F*** – Annexes A through F contain helpful information that is not mandatory.
  - ***Annex A*** contains a list of product safety standards. These standards provide further references for requirements that are in addition to the NEC® requirements for the electrical components mentioned.
  - ***Annex B*** contains information for determining ampacities of conductors under engineering supervision.
  - ***Annex C*** contains the conduit fill tables for multiple conductors of the same size and type within the accepted raceways.
  - ***Annex D*** contains examples of calculations for branch circuits, feeders, and services as well as other loads such as motor circuits.
  - ***Annex E*** contains information on types of building construction.
  - ***Annex F*** contains a cross-reference index for the articles that were renumbered in the latest edition of the code.

6.4.0 Text in the NEC®

When you open the NEC®, you will notice several different types of text or printing used. Here is an explanation of each type of text:

- Headings are in bold black letters.
- Italics identify exceptions. Exceptions explain the circumstances under which a specific part of the NEC® does not apply. Exceptions are written in italics under that part of the NEC® to which they pertain.
- **Fine print notes (FPNs)** explain something in an application, suggest other sections to read about the application, or provide tips about the application. These are defined in the text by the term (*FPN*) shown in parentheses before a paragraph in smaller print.

- Figures may be included with explanations to give you a picture of what your application may look like.
- Tables are often included when there is more than one possible application of the NEC®. You would use a table to look up the specifications of your application.

6.5.0 Navigating the NEC®

To locate information for a particular procedure, use the following steps:

Step 1 Familiarize yourself with ***NEC Articles 90, 100, and 110***, to gain an understanding of the material covered in the NEC® and the definitions used in it.

Step 2 Turn to the table of contents at the beginning of the NEC®.

Step 3 Locate the chapter that focuses on the desired category.

Step 4 Find the article pertaining to your specific application.

Step 5 Turn to the page indicated. Each application will begin with a bold heading.

> **NOTE**
> An index is provided at the end of the NEC®. The index lists specific topics and provides a reference to the location of the material within the NEC®. The index is helpful when you are looking for a specific topic rather than a general category.

Once you are familiar with ***NEC Articles 90, 100, and 110***, you can move on to the rest of the NEC®. There are several key sections used often in servicing low-voltage systems. *Table 1* summarizes these sections.

6.5.1 Example of Navigating the NEC®

Suppose you are installing a communications system cable at a building entrance. You know that the metallic members of the cable sheath need to be grounded, but you are not sure of the requirements. To find out this information, use the following procedure:

Step 1 Look in the NEC® table of contents and read down the list until you find an appropriate category. Since you are installing communications system cables, you would probably first look for the information you need in ***NEC Chapter 8***, *Communications Systems*. Further examination of the table of contents for ***NEC Chapter 8*** shows that ***NEC Article 800, Part IV*** covers grounding methods.

Step 2 If the information you are looking for is not readily identified in the table of contents, look in the NEC® index. Follow down the alphabetical listing of subjects until you find an appropriate category for our example, *Grounding*. Listed under *Grounding* you will find the entry *Communication Systems* with references to ***NEC Section 800.33*** and ***NEC Article 800, Part IV***.

Step 3 Referring to ***NEC Article 800, Part IV***, you will see that ***NEC Section 800.40*** covers cable and protector grounding.

Upon reading ***NEC Section 800.40*** and referenced ***NEC Section 800.33***, you know that the metallic members of the cable sheath must be grounded as close as practicable to the point of entrance or they must be interrupted as close to the point of entrance as practicable by an insulating joint or equivalent device. Furthermore, the metallic members of the cable sheath must be grounded using a grounding conductor and electrode that is connected via a bonding jumper to the building or structure power grounding electrode system. These connections must be made as specified in ***NEC Section 800.40 (A)*** through ***(D)***.

In addition, the connection made between the communications grounding conductor and electrode shall be made as specified in the applicable sections of ***NEC Article 250***.

7.0.0 ◆ *CANADIAN ELECTRICAL CODE, PART 1*

The Canadian Standards Association (CSA) sponsors, controls, and publishes the *Canadian Electrical Code, Part 1 (CE Code, Part 1)*. Like the NEC®, the intent of this code is to establish safety standards for the installation and maintenance of electrical equipment. As with the NEC®, the CE Code may be adopted and enforced by the provincial and territorial regulatory authorities.

Telecommunication and other low-voltage system installers who perform work in Canada must be familiar with the CE Code and comply with its requirements. *Table 2* summarizes several key sections of the CE Code used often in servicing low-voltage systems.

Table 1 Portions of the NEC® Affecting Telecommunications and Life Safety Systems (1 of 2)

NEC® Reference	Title	Description
NEC Section 90.2	Scope	The Scope provides information about what is covered in the NEC®. This section offers reference to the *National Electrical Safety Code®* (*NESC®*) for industrial or multi-building complexes.
NEC Section 90.3	Code Arrangement	This section explains how the NEC® chapters are positioned. Specifically, ***NEC Chapter 8, Communications Systems***, is an independent chapter except where reference is made to other chapters.
NEC Article 100, Part I	Definitions	Definitions are those not commonly defined in English dictionaries. Some terms of interest include *accessible, bonding, explosion-proof apparatus, ground, premises wiring,* and *signaling circuit.*
NEC Section 110.26	Working Space About Electric Equipment (600V, Nominal, or Less)	This section explains the space for working clearances around electrical equipment. This information is useful when placing a terminal in an electrical closet or electronic components on a communications rack.
NEC Article 250	Grounding	This article is referenced from ***NEC Article 800.*** It contains specific requirements for the communications grounding and bonding network.
NEC Section 300.21	Spread of Fire or Products of Combustion	This section is referenced from various NEC® articles in reference to abandoned cable.
NEC Section 300.22	Wiring in Ducts, Plenums, and Other Air Handling Spaces	This section is referenced from ***NEC Article 800 (NEC Section 800.53)*** and ***NEC Article 820 (NEC Section 820.3).*** This article covers the application of listed communication wires, cables, and raceways in ducts, plenums, and other air handling spaces.
NEC Article 500	Hazardous (Classified) Locations	All of ***NEC Article 500*** is referenced in ***NEC Article 800 (NEC Section 800.8).*** This article covers hazardous locations such as gasoline stations and industrial complexes. Additionally, healthcare facilities ***(NEC Section 517.80)*** are of particular importance. Theaters and marinas are also included in this article.
NEC Article 640	Audio Signal Processing, Amplification, and Reproduction Equipment	This article covers audio signal processing, amplification, and reproduction equipment.
NEC Article 725	Class 1, Class 2, and Class 3 Remote-Control, Signaling, and Power-Limited Systems	***NEC Article 725*** specifies circuits other than those used specifically for electrical light and power.
NEC Article 760	Fire Alarm Systems	***NEC Article 760*** contains requirements for the wiring and equipment used in fire alarm systems.

209T01A.EPS

Table 1 Portions of the NEC® Affecting Telecommunications and Life Safety Systems (2 of 2)

NEC® Reference	Title	Description
NEC Article 770	Optical Fiber Cables and Raceways	***NEC Article 770*** pertains to optical fiber cables and raceways. Within this section are the requirements for listing of cable, marking, and installation.
NEC Section 780.6	Cables and Conductors	Referenced from ***NEC Article 800 (NEC Section 800.3)***, this section contains requirements for listed hybrid cables used in closed loop and programmed power distribution systems.
NEC Article 800	Communications Systems	***NEC Article 800*** contains the requirements for communications circuits.
NEC Article 810	Radio and Television Equipment	***NEC Article 810*** contains the requirements for radio and television.
NEC Article 820	Community Antenna Television and Radio Distribution Systems	***NEC Article 820*** contains requirements for community antenna television and radio distribution systems.
NEC Article 830	Network-Powered Broadband Communication Circuits	***NEC Article 830*** contains the requirements for network-powered broadband communication circuits.

209T01B.EPS

Table 2 Portions of the CE Code Affecting Telecommunications and Life Safety Systems

CE Code Reference	Title	Description
2	General Rules	Provides information on the following: • Permits • Marking of cables • Flame spread requirements for electrical wiring and cables
10	Grounding and Bonding	Contains detailed grounding and bonding information and requirements for using and identifying grounding and bonding conductors
12	Wiring Methods	Involves the requirements for installing wiring systems and outlines the following: • Raceway systems • Boxes • Other system elements
56	Optical Fiber Cables	Contains the requirements for installing optical fiber cables
60	Electrical Communication Systems	Contains the requirements for installing communications circuits

209T02.EPS

8.0.0 ◆ NATIONAL FIRE ALARM CODE® (NFPA 72)

NFPA 72 covers the application, installation, performance, and maintenance of fire alarm systems and their components. The code defines the means of signal initiation, transmission, notification, and annunciation; the levels of performance; and the reliability of the various types of fire alarm systems. It defines the features associated with these systems and also provides the information necessary to modify or upgrade an existing system to meet the requirements of a particular system classification. The intent of the code is simply to define requirements; it is not intended to establish the methods by which the requirements are achieved. The first three chapters contain general information, including administration, referenced publications, and definitions. The remaining eight chapters contain detailed information covering fire alarm systems and their components:

- *Chapter 4, Fundamentals of Fire Alarm Systems* – This chapter covers the fire alarms systems, equipment, and components addressed in the remaining chapters, with the exception of household fire alarm systems, which are covered in Chapter 11.
- *Chapter 5, Initiating Devices* – This chapter defines the requirements for smoke, heat, and fire detectors, as well as alarm initiating devices.
- *Chapter 6, Protected Premises Fire Alarm Systems* – This chapter covers systems, other than household fire protection systems, that are used to protect life and property by detecting and annunciating the existence of heat, fire, or smoke within the protected premises.
- *Chapter 7, Notification Appliances for Fire Alarm Systems* – This chapter contains the requirements for audible and visible devices used to signal a fire alarm.
- *Chapter 8, Supervising Station Fire Alarm Systems* – This chapter covers the requirements for facilities, operation and service, and communications methods used in monitoring protected premises.
- *Chapter 9, Public Fire Alarm Reporting Systems* – This chapter addresses the installation and use of municipal fire alarm systems, including fireboxes and other systems used to initiate fire alarms from the street.
- *Chapter 10, Inspection, Maintenance, and Testing* – This chapter addresses testing and maintenance requirements for protected premises fire alarm systems, supervising station fire alarm reporting systems, public fire alarm reporting systems, and the initiating devices and notification appliances connected to them.
- *Chapter 11, Single- and Multiple-Station Alarms and Household Fire Alarm Systems* – This chapter covers all fire warning equipment used in residential applications.

The code also contains annexes that provide relevant explanatory material, an engineering guide for automatic fire detector spacing, and referenced publications.

9.0.0 ◆ LIFE SAFETY CODE® (NFPA 101®)

NFPA 101® addresses life safety from fire as it applies to both new and existing buildings. It covers those construction, protection, and occupancy features necessary to minimize danger to life from fire, including smoke, fumes, or panic. It identifies the minimum requirements for the design of exit facilities that will permit prompt escape of occupants from buildings or, where desirable, into safe areas within buildings. This code recognizes that life safety is more than a matter of egress and, accordingly, deals with other considerations that are essential to life safety. It does not address all of the general fire prevention or building construction features that are normally a function of fire prevention and building codes.

NFPA 101® is not adopted in all states or local jurisdictions. If it is adopted, there could be conflicts between the various codes and other adopted documents. The AHJ will determine which code takes precedence, although it is usually the most stringent.

The code is divided into 42 chapters. The first 10 chapters contain general information related to the code, as well as material related to building egress and fire safety, including the following:

- General information
- Fundamental requirements
- Definitions
- Classification of occupancy and hazard contents
- Means of egress
- Features of fire protection
- Building service and fire protection equipment

The remaining chapters cover different types of occupancies. The requirements given in these chapters apply to both new and existing occupancies. These chapters cover the requirements for the following types of occupancies:

- Assembly
- Educational
- Healthcare
- Detention and correctional
- Motels and dormitories
- Apartment buildings
- Lodging and rooming houses
- One- and two-family dwellings
- Residential board and care
- Mercantile
- Business
- Industrial
- Storage
- Daycare

Code annexes contain useful explanatory notes and references related to the text contained in the various chapters of the code.

10.0.0 ◆ RELATED NFPA CODES

Other codes published by NFPA that apply to low-voltage installations are briefly described in this section. These codes may be cited in project specifications; therefore, all installers should be familiar with their content.

10.1.0 *Standard for the Installation of Sprinkler Systems* (*NFPA 13*)

NFPA 13 covers the proper design and installation of sprinkler systems for all types of fire hazards. Coverage includes the following items:

- Design considerations
- Water supplies
- Equipment requirements
- System flow rates
- Sprinkler location and position
- General storage and rack storage of materials
- Specifically identified criteria for special occupancy hazards
- Minimum sizes for sprinklers used in storage applications
- Separation requirements between early suppression fast response and other types of sprinkler systems
- Rules for protecting sprinkler systems against seismic events.

Related standard *NFPA 13D* covers the installation of sprinkler systems in one- and two-family dwellings and manufactured homes. *NFPA 13R* covers the installation of sprinkler systems in residential occupancies up to and including four stories in height.

The fire sprinkler standards, and others (such as *NFPA 20, Standard for the Installation of Stationary Fire Pumps for Fire Protection*) contain requirements for alarms and signaling. Where a structure also requires a fire alarm system in accordance with *NFPA 72*, the more stringent provisions within each standard must be applied in the event of a conflict.

10.2.0 *Standard for the Protection of Information Technology Equipment* (*NFPA 75*)

NFPA 75 covers the requirements for computer installations needing fire protection and special building construction, rooms, areas, or operating environments. Application is based on risk considerations such as the business interruption aspects of the function, as in computers used in the stock market, or the fire threat to the installation.

10.3.0 *Standard for the Installation of Lightning Protection Systems* (*NFPA 780*)

NFPA 780 describes the requirements for the protection of people, buildings, special occupancies, heavy-duty stacks, structures containing flammable liquids and gases, and other entities against lightning damage.

11.0.0 ◆ NATIONAL BUILDING CODES

Building codes that are national in scope provide minimum standards to guard the life and safety of the public by regulating and controlling the design, construction, and quality of materials used in modern construction. They have also come to govern the use and occupancy, location of a type of building, and the on-going maintenance of all buildings and facilities. Once adopted by a local jurisdiction, these national building codes then become law. It is common for localities to change or add new requirements to any national code requirements adopted in order to meet more stringent requirements and/or local needs. The provisions of the national building codes apply to the construction, alteration, movement, demolition, repair, structural maintenance, and use of any building or structure within the local jurisdiction.

The national building codes are the legal instruments that enforce public safety in construction of human habitation and assembly structures. They are used not only in the construction industry but also by the insurance industry for compensation appraisals and claims adjustments, and by the legal industry for court litigation.

Up until 2000, there were three model building codes. The three code writing groups, Building Officials and Code Administrators (BOCA), International Conference of Building Officials (ICBO), and Southern Building Code Congress International (SBCCI), combined into one organization called the International Code Council (ICC) with the purpose of writing one nationally accepted family of building and fire codes. The first edition of the *International Building Code®* was published in 2000 and the second edition in 2003. It is intended to continue on a three-year cycle.

In 2002, the NFPA published its own building code, *NFPA 5000®*. There are now two nationally recognized codes competing for adoption by the 50 states.

The format and chapter organization of the two codes differ, but the content and subjects covered are generally the same. Both codes cover all types of occupancies from single-family residences to high-rise office buildings, as well as industrial facilities. They also cover structures, building materials, and building systems, including life safety systems.

Some major cities and local jurisdictions will have their own independent codes. As an EST, you must know the requirements within the jurisdiction where the work is to be performed.

12.0.0 ◆ ANSI/TIA/EIA TELECOMMUNICATIONS RELATED STANDARDS

The American National Standards Institute (ANSI), **Electronic Industries Alliance (EIA)**, and **Telecommunications Industry Association (TIA)** jointly publish standards for the manufacturing, installation, and performance of voice, data, video, audio, security, environmental control, and other electronic and communications equipment and systems. Five ANSI/TIA/EIA standards listed here govern telecommunications cabling in buildings. Each standard covers a specific part of building cabling. They address the required cable, hardware, equipment, design, and installation practices. In addition, each ANSI/TIA/EIA standard lists related standards and other reference materials that deal with the same topics.

- *ANSI/TIA/EIA-568-B, Commercial Building Telecommunications Cabling Standard*
- *ANSI/TIA/EIA-569-B, Commercial Building Standard for Telecommunications Pathways and Spaces*
- *ANSI/TIA/EIA-570-B, Residential Telecommunications Infrastructure*
- *ANSI/TIA/EIA-606, Administration Standard for the Telecommunications Infrastructure of Commercial Buildings*
- *ANSI/TIA/EIA-607, Commercial Building Grounding and Bonding Requirements for Telecommunications*
- *ANSI/TIA/EIA-942, Telecommunications Infrastructure Standard for Data Centers*

12.1.0 Commercial Building Telecommunications Cabling Standard (*ANSI/TIA/EIA-568-B*)

ANSI/TIA/EIA-568-B covers telecommunications cabling in commercial buildings. The standard provides specifications for a generic building cabling system that can be created and used with a variety of products from many different manufacturers. In addition to design specifications, the standard includes performance specifications for cables and components used in commercial building cabling.

The B revision to *ANSI/TIA/EIA-568* was published in 2001. This revision, in addition to incorporating outstanding technical service bulletins, addenda, and interim standards that were added to the previous edition (*568-A*), resulted in a major

INSIDE TRACK

Association Origins

Both TIA and EIA originated in 1924. TIA started as a small group organizing a telephone industry trade show. That group later became a committee of the United States Independent Telephone Association. It remained that way until 1979, when it split off as a separate association—The United States Telecommunications Suppliers Association (USTSA)—to organize telecom exhibits and trade shows. TIA was formed in 1988 by merging USTSA with the EIA Information and Telecommunications Technical Group. EIA originated in 1924 as the Radio Manufacturers Association.

Source: *Telecommunications Industry Association*

reorganization of the standard into three separate technical standards.

- *ANSI/TIA/EIA-568-B.1, General Requirements* – This section covers commercial building cabling for telecommunications products and services. It specifies a telecommunications cabling system for commercial buildings that will support a multi-product, multi-vendor environment.
- *ANSI/TIA/EIA-568-B.2, 100-Ohm Balanced Twisted-Pair Cabling Standard* – This standard covers copper components. It specifies cabling components, transmissions, and cabling models, as well as the measurement procedures needed for verification of balanced twisted pair cabling.
- *ANSI/TIA/EIA-568-B.3, Optical Fiber Cabling Components Standard* – This standard covers fiber components. It specifies components such as cables and connectors, along with the transmission requirements, for an optical fiber cabling system.

A number of addenda have been added since the 2001 publication of the standard. These include the following addenda, which had been issued by the end of 2004.

- *ANSI/TIA/EIA-568-B.1-1* – This addendum covers bend radii for 4-pair UTP and 4-pair ScTP patch cables.
- *ANSI/TIA/EIA-568-B.1-2* – This addendum provides additional requirements for grounding and bonding of installed balanced ScTP cables and hardware in commercial buildings.
- *ANSI/TIA/EIA-568-B.1-3* – This addendum applies to supportable distances and channel attenuation for optical fiber cable.
- *ANSI/TIA/EIA-568-B.1-4* – This addendum provides recognition for Category 6 and 850 nm laser-optimized 50/125 µm multimode optical fiber cable.
- *ANSI/TIA/EIA-568-B.1-5* – This addendum covers cabling requirements for telecommunications enclosures.
- *ANSI/TIA/EIA-568-B.2-1* – This addendum specifies requirements for 100-ohm Category 6 cables and connectors.
- *ANSI/TIA/EIA-568-B.2-2* – This addendum makes corrections to *ANSI/TIA/EIA-568-B.2.*
- *ANSI/TIA/EIA-568-B.2-3* – This addendum provides additional considerations for insertion loss and return loss pass/fail criteria.
- *ANSI/TIA/EIA-568-B.2-4* – This addendum specifies solderless connection reliability for copper connections used in commercial buildings.
- *ANSI/TIA/EIA-568-B.2-5* – This addendum corrects references in *ANSI/TIA/EIA-568-B.2.*
- *ANSI/TIA/EIA-568-B.2-6* – This addendum refines and enhances measurement methods specified for Category 6 cable.
- *ANSI/TIA/EIA-568-B.3-1* – This addendum provides transmission performance standards for 50/125 µm optical fiber cables.

12.2.0 *Commercial Building Standard for Telecommunications Pathways and Spaces (ANSI/TIA/EIA-569-B)*

ANSI/TIA/EIA-569-B provides specifications for designing and constructing the pathways and spaces for telecommunications cabling in commercial buildings. The standard is especially useful for writing construction bids and contracts.

The standard's specifications are based on the belief that the building, the telecommunications needs of its occupants, and the telecommunications technology available may change several times during the life of a building. This standard covers all low-voltage systems that carry information inside a building, including telecommunications pathways and spaces.

ANSI/TIA/EIA-569-B was released in 2004. It has the following structure:

1. *Scope*
2. *Definition of Terms, Acronyms and Abbreviations, Units of Measure, and Symbols*
3. *Telecommunications Diversity*
4. *Entrance Facilities*
5. *Access Provider Spaces and Service Provider Spaces*
6. *Multi-Tenant Building Spaces*

INSIDE TRACK

Using Up-to-Date Standards

If your work requires you to refer to standards, you must make sure that you have access to the current version of the standard as well as any addenda to the standard. Addenda, technical service bulletins, and interim standards supplement, clarify, and correct the basic standard. Look through the list of addenda to *ANSI/TIA/EIA-568-B.* You should be able to see how the content of these addenda might have an impact on telecommunications installations.

7. *Building Spaces*
8. *Tenant Building Pathways*

In addition to setting specifications, the standard offers a good working definition of each type of space or pathway discussed. The standard also contains many illustrations of pathways and equipment.

ANSI/TIA/EIA-569-B also includes four annexes:

- *Annex A – Firestopping*
- *Annex B – Additional Section Information*
- *Annex C – Noise Reduction Guidelines*
- *Annex D – Bibliography and References*

12.3.0 *Residential Telecommunications Infrastructure (ANSI/TIA/EIA-570-B)*

ANSI/TIA/EIA-570-B provides standardized requirements for residential telecommunications cabling. The standard's specifications are based on the facilities that are necessary for existing and emerging telecommunications services. Within the standard, services are correlated to grades of cabling for residential units. The cabling requirements covered include support for voice, data, and video; security systems; whole-home audio cabling; and control systems. Cable and connector requirements cover UTP cabling; 75-ohm coaxial cabling; optical fiber cabling; and multi-conductor cabling. Installation requirements are provided for all these types as well as special cabling for security, control, and audio systems. Revision B of the standard was released in 2004. It contains eight sections:

1. *Scope*
2. *Definition of Terms, Acronyms and Abbreviations, Units of Measure, and Symbols*
3. *Single-Dwelling Residence Infrastructure*
4. *Multi-Dwelling/Campus Infrastructure*
5. *Cable and Connecting Hardware*
6. *Installation Requirements*
7. *Field Test Requirements*
8. *Administration*

ANSI/TIA/EIA-570-B also includes four informative annexes:

- *Annex A – Cabling Residential Buildings*
- *Annex B – Installation Guide*
- *Annex C – Typical Applications that Interface to Residential Cabling*
- *Annex D – Bibliography and References*

12.4.0 *Administration Standard for the Telecommunications Infrastructure of Commercial Buildings (ANSI/TIA/EIA-606)*

ANSI/TIA/EIA-606 describes the requirements for the keeping of records and the information that must be available in order to administer the telecommunications system in a commercial building properly. To ensure that administration records are accurate and up-to-date, cabling installers must understand the following concepts:

- What information the records must contain
- How the information should be recorded
- What devices and structures must be labeled
- How the devices and structures should be labeled

The standard requires that an administration system track all aspects of the telecommunications system, including cables, terminations, termination hardware, patching and cross-connect facilities, splices, conduit and other pathways, bonding and grounding, telecommunications rooms, equipment rooms, and other telecommunications spaces.

Each component of the telecommunications system must have a unique identification code. This code is used for the component's record, for the record of connected or related components, and for any drawings where the components appear. If the component requires labeling, the code must be printed on the component's label or on the component itself. Details about identification codes appear in *Section 4.2, Identifiers.*

Section 4.3, Records explains what information must be recorded. Examples of records appear throughout the standard. These examples show both the required information and optional information that would make the record more complete and provide linkage between the components.

Although not part of the standard, *Appendix C* contains a list of common telecommunications symbols used in representing elements of the telecommunications infrastructure.

12.5.0 *Commercial Building Grounding and Bonding Requirements for Telecommunications (ANSI/TIA/EIA-607)*

ANSI/TIA/EIA-607 specifies the grounding and bonding requirements for the telecommunications system in a commercial building. It also specifies the interconnections between telecommunications grounding and other grounding. This standard specifies the following:

- Ground reference for telecommunications systems in entrance facilities, telecommunications rooms, and equipment rooms
- Bonding and connecting of conductors, cable shields, pathways, and hardware in entrance facilities, telecommunications rooms, and equipment rooms

ANSI/TIA/EIA-607 also describes in detail the physical requirements for the five major components of a telecommunications grounding and bonding system:

- Telecommunications bonding backbone (TBB)
- Telecommunications main grounding busbar (TMGB)
- Telecommunications grounding busbar (TGB)
- Telecommunications bonding backbone interconnecting bonding conductor (TBBIBC)
- Bonding conductor for telecommunications

The interconnections between these components and their relationship to other building grounds are explained in detail in the standard. The explanations include installation considerations.

It is specified that all bonding conductors must be a 6 AWG or larger insulated copper wire marked by a distinctive green color. The main bonding conductor must also be labeled with a warning, which asks that any disconnection of this bonding conductor be reported immediately to the building's telecommunications manager.

13.0.0 ◆ RELATED STANDARDS

There are related standards applicable to low-voltage systems published by the **Institute of Electrical and Electronic Engineers (IEEE)**. Some commonly used standards are briefly described here.

13.1.0 *CSMA/CD Access Method (IEEE Standard 802.3)*

This standard covers the telecommunications and information exchange between systems and local and metropolitan area networks. Covered are specific requirements for carrier sense multiple access with collision detection (CSMA/CD) access methods and physical layer specifications. Several supplements to this specification are also available to cover the following subjects: frame extensions for virtual bridged local area networks; physical layer parameters and specifications for 1000MB/s operation over four-pair of Category 5 balanced copper cabling, type 1000BASE-T; and link aggregation.

13.2.0 *Token Ring Access Method (IEEE Standard 802.5)*

This standard covers the telecommunications and information exchange between systems and local and metropolitan area networks. Covered are specific requirements for token ring access methods and physical layer specifications. Supplements to this specification are also available to cover the following subjects: recommended practice for dual ring operation with wrapback reconfiguration and 100Mbit/s dedicated ring operation over two-pair cabling.

13.3.0 *High-Performance Serial Bus (IEEE Standard 1394)*

This specification covers the requirements for the integration of a high-speed serial bus with IEEE standard 32-bit and 64-bit parallel buses, as well as such nonbus interconnects as the *IEEE Standard 1596*, scalable coherent interface. It is intended to provide a low-cost interconnect between cards on the same backplane, cards on other backplanes, and external peripherals. Specifications covered include summary descriptions, cable specifications, backplane specifications, link layer specifications, transaction layer specifications, and serial bus management specifications.

14.0.0 ◆ TESTING LABORATORIES

Testing laboratories are an integral part of the development of codes and standards. The NFPA and other organizations rely on testing laboratories to conduct research into electrical equipment and its safety. These laboratories perform extensive testing of new products to make sure that they are built to NEC® standards for electrical and fire safety. They receive statistics and reports from agencies all over the United States concerning electrical shocks and fires and their causes. Upon seeing developing trends concerning the association of certain equipment and dangerous situations or circumstances, this equipment is specifically targeted for research. All the reports from these laboratories are used in the generation of changes or revisions to the NEC® and other codes.

INSIDE TRACK

Underwriters Laboratories, Inc.

The Chicago World's Fair was opened in 1893 and thanks to Edison's introduction of the electric light bulb, the World's Fair lit up the world. But all was not perfect—wires soon sputtered and crackled, and, ironically, the Palace of Electricity caught fire. The fair's insurance company called in a troubleshooting engineer, who, after careful inspection, found faulty and brittle insulation, worn out and deteriorated wiring, bare wires, and overloaded circuits.

He called for standards in the electrical industry, and then set up a testing laboratory above a Chicago firehouse to do just that. Hence, Underwriters Laboratories, Inc., an independent testing organization, was born. Underwriters Laboratories, Inc. (UL), is now an internationally recognized authority on product safety testing and safety certification and standards development.

14.1.0 Nationally Recognized Testing Laboratories

Nationally Recognized Testing Laboratories (NRTLs) are product safety certification laboratories. They establish and operate product safety certification programs to make sure that items produced under the service are safeguarded against reasonably foreseeable risks. NRTLs maintain a worldwide network of field representatives who make unannounced visits to factories to check products bearing their safety marks.

Two such nationally recognized testing laboratories are the UL and the Canadian Standards Association (CSA). The UL is a not-for-profit organization whose principle business is to evaluate U.S. products to determine their compliance with defined standards for safety. UL evaluates electrical and mechanical products, building materials, construction systems, fire protection equipment, and marine products. UL's standards for safety establish the basis for testing products and levels of acceptability. Once a product has successfully passed the testing process, it may be listed and have the UL label affixed to it. The CSA performs product evaluations and tests on Canadian products similar to those performed by the UL. Through international agreements, both organizations are authorized to test and certify specific types of products for use in both U.S. and Canadian markets.

14.2.0 National Electrical Manufacturers Association

The National Electrical Manufacturers Association (NEMA) was founded in 1926. It is made up of companies that manufacture equipment used for generation, transmission, distribution, control, and utilization of electric power. The objectives of NEMA are to maintain and improve the quality and reliability of products; to ensure safety standards in the manufacture and use of products; and to develop product standards covering such matters as naming, rating, performance, testing, and dimensions. NEMA participates in developing the NEC® and the *National Electrical Safety Code®* and advocates their acceptance by state and local authorities.

Summary

In addition to learning the technical aspects of your profession, you must become familiar with the codes and standards that govern your work. Most of the work done by ESTs in telecommunications, life safety, security, and other systems is governed by industry standards and building codes. The NEC® specifies the minimum provisions necessary for protecting people and property from hazards arising from the use of electricity and electrical equipment. As an EST, you must be aware of how to use and apply the NEC® on your job. Using the NEC® correctly will help you to install and maintain the electrical equipment and systems you will encounter.

As an EST, you will be the last person on the design/installation team to see that systems are installed correctly and the devices operate as intended. Your knowledge of the relevant codes and standards, and their requirements, will enable you and your company to provide the customer with an approved, high-quality system.

Review Questions

1. The NEC® is published by the _____.
   a. National Fire Protection Association
   b. National Electrical Code Association
   c. International Code Council
   d. American National Standards Institute

2. A mandatory rule contains the words _____ and must be adhered to.
   a. shall or shall not
   b. accepted or not accepted
   c. should or should not
   d. approved or not approved

3. If there is a conflict between two codes or standards that apply to a job, the best approach is to _____
   a. apply the most demanding standard
   b. apply the least demanding standard
   c. call the organizations that publish the codes
   d. use the average of the two requirements from both standards

4. The NEC® provides the _____ requirements for the installation of electrical systems.
   a. most stringent
   b. complete
   c. design specification
   d. minimum

5. *NEC Article 110* covers _____.
   a. branch circuits
   b. definitions
   c. general requirements for electrical installations
   d. wiring design and protection

6. Hazardous locations are covered in _____.
   a. ***NEC Article 250***
   b. ***NEC Article 500***
   c. ***NEC Article 725***
   d. ***NEC Article 760***

7. Which of the following codes is the *National Fire Alarm Code®*?
   a. *NFPA 13*
   b. *NFPA 72*
   c. *NFPA 75*
   d. *NFPA 101*

8. Regarding building codes, _____.
   a. national codes always take precedence over local codes
   b. local codes are the same as national codes, so there is no conflict
   c. local codes may contain requirements that are not covered in national codes
   d. they are for information only and do not have to be rigidly followed

9. The ANSI/TIA/EIA standard that governs commercial pathways and spaces is _____.
   a. *ANSI/TIA/EIA-568-B*
   b. *ANSI/TIA/EIA-569-B*
   c. *ANSI/TIA/EIA-570-B*
   d. *ANSI/TIA/EIA-606*

10. The function of Underwriters Laboratories, Inc., is to_____.
    a. ensure electronic systems against damage or failure
    b. check installed systems to make sure they work correctly
    c. develop product standards for electronic systems and equipment
    d. conduct research into electrical equipment and its safety

GLOSSARY

Trade Terms Introduced in This Module

American National Standards Institute (ANSI): An international organization that serves in the capacity as administrator and coordinator of the U.S. private sector voluntary standardization system. ANSI does not itself develop standards, rather it facilitates development by establishing consensus among qualified groups.

Electronic Industries Alliance (EIA): An international organization whose goal is to enhance the competitiveness of American electronics producers. The organization supports functions for its members in areas of market statistics, technical standards, government relations, and public affairs.

Institute of Electrical and Electronic Engineers (IEEE): An international organization that produces about 30 percent of the world's published literature and standards pertaining to electrical engineering, computers, and control technology.

National Electrical Manufacturers Association (NEMA): A group made up of companies that manufacture equipment used for the generation, transmission, distribution, control, and utilization of electric power. The objectives of NEMA are to maintain and improve the quality and reliability of products; to ensure safety standards in the manufacture and use of products; and to develop product standards covering such matters as naming, rating, performance, testing, and dimensions.

National Fire Protection Association (NFPA): A nonprofit organization that publishes codes and standards that are nationally recognized by state and local governments as adoptable into law. Some of these codes and standards include the *National Electrical Code®*, *Life Safety Code®*, *Uniform Fire Code™*, *National Fuel Gas Code*, *National Fire Alarm Code®*, and the *Building and Construction Safety Code™*. The organization's mission is to reduce the burden of fire and other hazards on the quality of life by providing and advocating codes and standards, research, training, and education.

Nationally Recognized Testing Laboratories (NRTL): A national group of product safety certification laboratories. They establish and operate product safety certification programs to ensure that items produced under the service are safeguarded against reasonably foreseeable risks.

Telecommunications Industry Association (TIA): An international trade organization that provides communications and information technology products, materials, systems, distribution services, and professional services. It represents suppliers of communications and information technology products on public policy, standards, and market-development issues.

Underwriters Laboratories (UL): A group of laboratories that tests devices and materials for compliance with the standards of construction and performance established by the laboratory and with regards to the suitability for installation in accordance with the appropriate standards of the National Board of Fire Underwriters.

REFERENCES & ACKNOWLEDGMENTS

Additional Resources

This module is intended to be a thorough resource for task training. The following reference works are suggested for further study. These are optional materials for continued education rather than for task training.

ANSI/TIA/EIA Standards, Latest Edition, available from Global Engineering Documents, Englewood, CO.

ANSI/TIA/EIA-568-B, Commercial Building Telecommunications Cabling Standard Plus Addenda 1 through 3, and Telecommunications Systems Bulletins TSB-67, TSB-72, and TSB-75

ANSI/TIA/EIA-569-B, Commercial Building Standards for Telecommunications Pathways and Spaces

ANSI/TIA/EIA-570-B, Residential Telecommunications Cabling Standard

ANSI/TIA/EIA-606, Administration Standard for the Telecommunications Infrastructure of Commercial Buildings

ANSI/TIA/EIA-607, Commercial Building Grounding and Bonding Requirements for Telecommunications

IEEE Standards, Latest Edition. Piscataway, NJ: Institute of Electrical and Electronic Engineers.

Standard 802.3, CSMA/CD Access Method

Standard 802.5, Token Ring Access Method

Standard 1394, Standard for a High Performance Serial Bus

International Building Code, Falls Church, VA: International Code Council.

NFPA Codes, Latest Edition. Quincy, MA: National Fire Protection Association.

NFPA 5000, Building Construction and Safety Code™

NFPA 13, Installation of Sprinkler Systems

NFPA 13D, Installation of Sprinkler Systems in One- and Two-Family Dwellings and Manufactured Homes

NFPA 13R, Installation of Sprinkler Systems in Residential Occupancies up to and Including Four Stories in Height

NFPA 70, National Electrical Code®

NFPA 72, National Fire Alarm Code®

NFPA 75, Protection of Information Technology Equipment

NFPA 77, Recommended Practice on Static Electricity

NFPA 101, Life Safety Code®

NFPA 780, Standard For Installation Of Lightning Protection Systems

SCTE Standards, Latest Edition. Exton, PA: Society of Cable Telecommunications Engineers.

ANSI/SCTE 74 2003 (formerly *IPS SP 001*), *Specification for Braided 75 Ohm Flexible RF Coaxial Drop Cable*

Websites:

www.bicsi.org

www.nfpa.org

www.tiaonline.org

Figure Credits

Tim Ely	209SA01

CONTREN® LEARNING SERIES — USER UPDATE

NCCER makes every effort to keep these textbooks up-to-date and free of technical errors. We appreciate your help in this process. If you have an idea for improving this textbook, or if you find an error, a typographical mistake, or an inaccuracy in NCCER's Contren® textbooks, please write us, using this form or a photocopy. Be sure to include the exact module number, page number, a detailed description, and the correction, if applicable. Your input will be brought to the attention of the Technical Review Committee. Thank you for your assistance.

Instructors – If you found that additional materials were necessary in order to teach this module effectively, please let us know so that we may include them in the Equipment/Materials list in the Annotated Instructor's Guide.

Write: Product Development and Revision
National Center for Construction Education and Research
P.O. Box 141104, Gainesville, FL 32614-1104

Fax: 352-334-0932

E-mail: curriculum@nccer.org

Craft ______________________ Module Name ______________________

Copyright Date __________ Module Number __________ Page Number(s) __________

Description

____

____

____

(Optional) Correction

____

____

(Optional) Your Name and Address

____

____